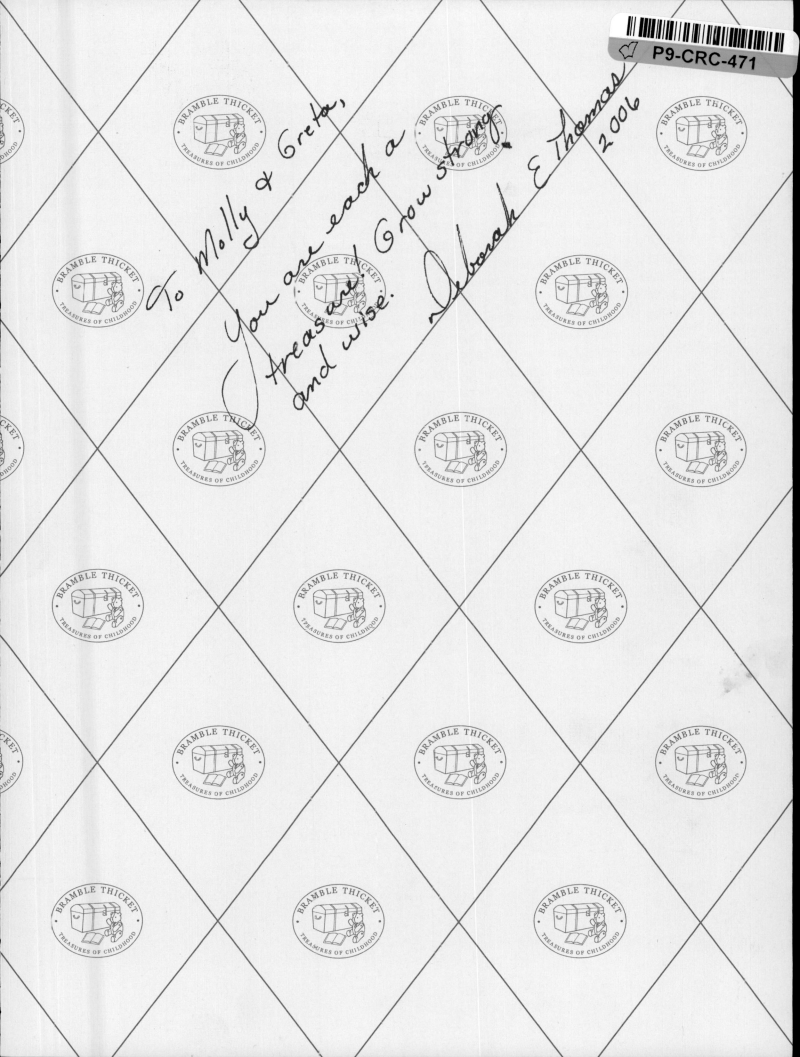

To Molly & Greta,

You are each a
Treasure. Grow strong
and wise.

Deborah E Thomas
2006

THE
BRAMBLE THICKET

Deborah Edge Thomas

A STORY FOR THOSE WHO LOVE ADVENTURE
AND SMALL BEARS

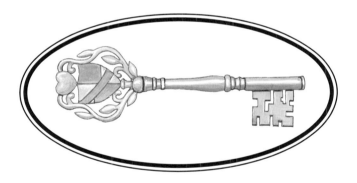

WITH ILLUSTRATIONS BY James E. Seward

First Edition
2003

To

James W. Thomas,

my husband,

a most extraordinary man.

A dreamer who believes

in the diligent, daily pursuit of excellence

through work and laughter;

he was the first to hear this story.

Illustrations by James Seward

Published by Gypsy Hill Publishing
Cookeville, Tennessee
Printed by Falcon Press
Prepress: Charles Lewis
Cover Design: Mike Lang
Book Design: Deborah E. Thomas
A Bramble Thicket Book

ISBN:0-9742805-0-X

Printed and bound in the United States of America

For speaking engagements, Deborah may be contacted at www.bramblethicket.com

Bramble™ Whisper™ Thicket™ Helper™ Rough™ Tumble™ Ramble™ Kandoo™ Candlelight™ Tiggles™ and Beartherapy™
Endsheets & Logo are trademarks of THE BRAMBLE THICKET COMPANY.

Library of Congress Control Number: 2003096518

Bramble

Whisper

Tumble

Table of Contents

Ramble

Kandoo

Thicket

Rough

Helper

CHAPTER 1

The Discovery

It was a damp, gray, drizzly morning. The drops of rain dripped off the leaves and ran over the white stones along the edge of the brook. Down the forest path came seven little bears each holding a Mayapple leaf as an umbrella.

This little group was a bright spot in the early morning fog. Yes, a bright spot, for each bear was a different color! The colors were so beautiful, it was as though a rainbow's light had painted them. As the morning sun peeked from behind a cloud, the wet gray of the forest sparkled like glittering diamonds!

"Come on, let's keep looking," spoke one bear. "We must find a new home and some berries for breakfast."

The others agreed. The seven little bears (with their umbrellas) began to search for berry bushes. A small brook flowing beside the path seemed to whisper, "Follow me to the berry bushes!" Following the brook, they soon discovered a fine patch of blueberries. The short, scrubby bushes made picking the berries easy for the short Bramble Thicket Bears!

A row of enormous (*enormous* means very, very big) fir trees stood guard on the other side of the brook.

"Let's explore under those trees," said a bright blue bear named Bramble. "We can cross the brook on these flat stones." And that is exactly what they did!

One by one they hopped across the brook. One by one they crawled under the low hanging fir branches. One, two, three, four, five, six, seven. They were glad to find that the ground was dry and covered with soft pine needles. Exploring is a lot of work! The tired bears decided to rest in the shelter of the boughs. They sat on the pine needles and listened to the pitter-patter of the raindrops. Not far away, the brook babbled as the water hurried to the sea.

Suddenly, the bears heard something else! They huddled close together. The noise was frightening!

The youngest of the bears, a bright yellow little fellow named Kandoo, crawled between Whisper and Bramble and crowded into the middle of the circle.

Bramble, being the oldest, listened very carefully . . . the noise seemed to come from the forest. It sounded like someone crying. Bramble asked the soft green twins, Rough and Tumble, to go with him to investigate. (*Investigate* means to learn or discover.) The three bears tip-toed around the giant tree trunk to the wall of evergreen boughs. Crawling forward, they peered from under the edge of the fir tree's branches. The tall grass of a meadow was all that they could see. Water dripped down on Rough's nose. He sneezed, "KACHOO!"

The crying stopped.

"What was that? Is anyone there?" asked a frightened voice.

The bears scrambled back out of sight. In a few moments, the sobbing began again. Rough looked like he was going to sneeze, so Bramble motioned to Tumble and Kandoo to come investigate once more. This time, he let Kandoo climb on his back, then Tumble climbed up on Kandoo's shoulders and looked through an opening in the branches. The other bears came close and began to ask questions.

"What do you see?"

"Who is making that noise?"

"Can we see too?"

Tumble spoke from where he balanced high on the shoulders of Kandoo, "Sshh! I never saw anything like that!"

Slowly, he began to climb down. Using Bramble's head for a step, he soon landed on the ground.

"Ouch," said Bramble, "don't step on my head!"

"Sorry, I wasn't thinking about you, I was thinking about her!" Tumble apologized.

"HER?" chorused the bears, "Who is 'her?'"

Tumble scratched his ear, (as he always does when he is trying to figure something out).

"'Her,' is a girl. At least I think it's a girl . . . but something is strange . . . she has wheels." He finished speaking and sat down very slowly.

"Wheels?" asked Thicket, a beautiful burgundy colored bear who loves ribbons, "Did you say, <u>wheels</u>?"

Tumble nodded, "I **definitely** saw wheels."

"Well," said Helper, a light blue bear, "wheels or not, it must be a girl and she is crying, so we must help her."

While the other bears talked, a lovely pink bear named Whisper noticed a large rock and climbed up on it. She peeked through the branches and listened very carefully. The top of the girl's head was all she could see.

The crying had stopped. It was very quiet. Whisper watched for another moment and then climbed down to join the others.

"She is sleeping," Whisper told them.

All the bears listened. There was no more crying. Once again they could hear the babbling water in the little brook.

Bramble was the first to wiggle out from beneath the fir tree to investigate. He hurried to a large log and hid behind it. His friends watched. They could hear the SPLAT, SPLAT, SPLAT of water as it dripped off the branches.

Rough was next. He ran and hid behind a berry bush. Helper wiggled out and dashed behind an old stump. Thicket and Whisper tip-toed to the log where Bramble hid. Tumble ran, did three-and-a-half somersaults, and landed in the middle of some very tall grass.

Bramble waved his paw in a circle. The bears moved forward very slowly and surrounded the strange girl with wheels. They were very close to her and spoke very quietly.

"Look! She has legs!" Thicket exclaimed.

"Yes, she does!" agreed Bramble.

Rough poked his brother as he said, "Tumble, it's the chair that has wheels, not the girl!"

"Oh," said Tumble, in a very small voice, as he hung his head and felt very sad, "It looked like a girl with wheels."

"Sshh, you'll wake her up," warned Bramble, "Where's Kandoo?"

"I'm here," said a voice from under a large green mound of Mayapple umbrellas.

"Where?" asked Whisper.

"Under here. I am 'camo-flagged,'" Kandoo spoke as he peered out from under the Mayapples. "Ramble showed me how to be 'camo-flagged.' That means hidden," Kandoo said with pride!

The older bears chuckled quietly.

"The word is camouflaged. It is one word," laughed Bramble as he removed a Mayapple from Kandoo's head. "You can come out now. She is sound asleep."

"I'm here," said a voice from under a mound of Mayapple umbrellas.

CHAPTER 2

Ramble's Cave

The dampness of the fog began to turn into a misty rain.

"We must get her to a dry place," Bramble said. "Let's take her back to the big fir tree."

The seven little bears had a very big job to do! They jabbered and whispered together as they made their plans. Helper, who was very good at thinking of how to do things, told the others his plan. Soon everyone was busy.

The sun came out and chased away the misty rain as the bears began to work. Bramble, Rough, and Tumble went into the forest to fetch some strong vines. Helper, Thicket, Whisper and Kandoo worked together to move sticks and stones out of the way.

Tumble came rolling down the hill with his vine wrapped all around himself. He was so tangled up that the other bears had to help untangle him!

"What are the vines for?" asked Kandoo.

"They will be our ropes," answered Helper as he tied a vine to the chair with wheels.

The girl continued to sleep. Soon the Bramble Thicket Bears were ready to move her to a dry place. Bramble, Whisper and Rough picked up the grape vine and put it over their shoulders.

Thicket and Tumble were ready to push the wheelchair, and Kandoo took his place at the right wheel. Helper supervised and encouraged everyone.

With the vine rope over their shoulders, three little bears pulled and tugged, tugged and pulled; while three more little bears Pushed and Pushed with all their might! The chair moved only a few inches. (Bramble Thicket Bears are very small, but they always do their very best.) Once more they Tugged, Pulled, and Pushed! Once more the chair moved only a few inches.

"Something is wrong," Bramble said as he sat down to rest.

"I know what is wrong," said Thicket quietly. "The wheels don't turn."

"DON'T TURN! THE WHEELS DON'T TURN?" shouted Helper, as he saw the deep ruts in the ground behind the chair.

"Who's there?" asked a frightened, sleepy voice.

"Oops," said Helper, as he looked up.

8

Looking down at him were the most beautiful blue eyes he had ever seen! Red-gold hair cascaded around a lovely face. Helper was awestruck; he couldn't say a word.

"Hello, who are you?" asked the girl, for it was she who had awakened.

"I'm . . . uhm . . . I'm," Helper shook his head.

Whisper spoke, "He's Helper and I'm Whisper. We are Bramble Thicket Bears. Who are you?"

The young lady answered, "I'm Candlelight, and I'm lost." Tears welled in her eyes as she tried to be brave.

Whisper introduced the other bears. Bramble bowed politely to Candlelight. Kandoo smiled shyly. Thicket offered a daisy and said hello. The twins, Rough and Tumble, said hello while standing on their heads. (Which is <u>not</u> Good Manners!)

"We found you! We are trying to take you to the big fir tree where it is dry," said Thicket as she looked at her wet ribbons.

"Your wheels don't turn," stated Kandoo as he pointed to the big wheels.

As she looked at the vine stretched before her, Candlelight could see the bears had been working hard.

"The wheels have a brake," she explained. "This handle sets the brake to stop the wheels." She showed the brake handle to her new friends.

"If I move the handle back, the brake is released, and the wheels can turn. All wheelchairs have brakes."

"Is this a wheelchair?" asked Rough.

"Yes, it is a wheelchair," Candlelight assured him.

The next few minutes were spent examining the wheelchair. Helper was quite fascinated! He liked to learn about new machines.

Once again everyone took their places. Candlelight released the brake. The bears Pulled, Tugged and Pushed! This time the wheels turned around and around as they moved toward the fir trees.

Small bears can go many places a young lady and a wheelchair cannot go. The bears discovered this when they tried to push the chair under the fir branches. Candlelight and the wheelchair were too big! What could they do? How could they find a dry place big enough for their new friend? The bears sat down to ponder the problem. (*Ponder* means to think very hard.)

Candlelight laughed when she saw them sitting on a nearby log. "You look like a rainbow," she exclaimed!

"We know," said Rough, "but looking like a rainbow does not help. We need Ramble, he knows a lot about the woods."

"He should be here soon," Bramble spoke hopefully. "He promised to meet us today here in this valley."

Just then the sound of whistling was heard; the notes seemed to float on the breeze. Everyone listened, but it was Kandoo who said, "That is Ramble's song! That is what he always whistles!"

Ramble himself soon appeared from behind some nearby huckleberry bushes.

"Hello, Kandoo! It's been a long time since I saw you!" Ramble shook paws with his young friend.

There was much hugging and pawshaking, for indeed, it had been a long time, and much had happened since the friends had been together. Candlelight was introduced; she was delighted to meet Ramble, a bright orange bear. She was fascinated to see his knapsack and whistle.

As soon as things settled down and Ramble understood the problem of finding a dry place for Candlelight, he spoke, "I know where we should go! Nearby is a place I call a cave, but it's not really a cave. It's just a rocky overhang along the path. We will be safe and dry there."

Once again, the bears picked up the vine and pulled the wheel-chair down the path. Candlelight helped by turning the wheels with her hands. It was a sight to see! They worked hard, for they were determined not to leave her alone. At last everyone was together underneath the rocky roof.

The evening shadows crept out of their hiding places as the bears gathered nuts and berries for supper. As the moon sent its silvery beams among the forest trees, the bears cuddled close to Candlelight.

"We will keep you warm," said Thicket.

Candlelight smiled at each of her new friends. "Thank you for finding me. We should call this place 'RAMBLE'S CAVE.'"

"That's a good idea," mumbled Ramble sleepily.

And that is exactly what they did!

The moon sent its silvery beams among the forest trees.

CHAPTER 3

Candlelight's Story

The morning sun painted the forest with rays of light. The sound of the brook filled the air like a song. Bluebirds singing in the trees seemed to say, "Cheer up! . . . Be of good cheer . . . cheer up!"

Thicket, who was a rather curious bear, climbed up onto the lap of her new friend.

"Why are you lost?" she asked.

Candlelight, who had been laughing at the antics of Rough and Tumble as they played together, became very quiet. She hugged Thicket and answered, "It is a sad story."

13

From under the wheelchair, Kandoo peered up at Candlelight. "Sharing your troubles makes things better," he announced.

The other bears gathered around Candlelight. They nodded their heads when Whisper said, "Friends can help, and we are friends."

Candlelight smiled through the tears that glistened in her eyes. Then she began to tell her story . . .

"I am an orphan," she said.

"I thought you were a girl. What's an orphan?" asked Kandoo.

"An orphan is a child who has no living parents," answered the girl. "Orphans are sometimes girls and sometimes boys."

"Oh," said Kandoo. "That is sad."

"Sshh," said the other bears.

Candlelight smiled as she continued her story, "I was away at school when I hurt my legs. The doctor told me I must have therapy, and now I live with my grandfather."

"What is 'there-a-pea?'" asked Kandoo.

"Therapy is very special exercises to help the body work again," Candlelight answered as she smiled at Kandoo.

The other bears were glad Kandoo had asked because they did not know what therapy was either.

"Does your grandfather live here in the woods?" asked Thicket.

14

"No, he lives in a beautiful stone house surrounded by lovely gardens. I came to live with him a few weeks ago. Grandfather is very kind and made me feel quite comfortable. Next week a nurse is coming to continue my therapy and help me with my lessons," Candlelight paused, "but I won't be there."

"How did you get HERE?" asked Tumble as he hung up-side down from a tree limb.

Candlelight continued, "Yesterday morning, Grandfather left very early on a business trip. Mrs. Pennywhistle, his housekeeper, had the day off. The maid was to take care of me. She was very unkind and complained about helping me. After breakfast, she called the chauffeur into the kitchen."

"What's a 'show-fer?'" interrupted Kandoo.

"Kandoo!" exclaimed Bramble. "Remember your manners! If you have a question, say 'excuse me,' or raise your paw!"

"Excuse me, what's a 'show-fer?'" asked Kandoo with his paw raised and his eye on Bramble.

"A chauffeur is a person who drives an automobile for someone else," Candlelight answered. "As I was saying, he and the maid came into my room and said we would be going for a ride. They whispered together and were in a great hurry. Soon we were in the car. We went up and down the hills and around many curves. I was lost and confused." She looked at the bears and sighed.

"At last the car turned down a very old, bumpy lane which led into the forest. Then we stopped. My chair was placed in the meadow and I was put into the chair. The chauffeur was very quiet, but the maid was nasty. She said, 'Someone will find you soon enough!' Then they hurried to the car and drove away."

The Bramble Thicket bears were shocked to learn this!

Candlelight continued, "I was very frightened. I tried to move my wheelchair back to the lane. The grass was so tall, I could not move it by myself. It started to rain. I began to cry. I cried myself to sleep."

"And we found you!" shouted Kandoo, as he danced a jig! "And you can live with us forever!"

Candlelight laughed! The little bear looked so funny, everyone had to laugh with her! "Thank you," she said, "but my grandfather and Mrs. Pennywhistle will be very worried. I must find a way to go home."

Helper stood with his paws on his hips. "Besides, we don't have a home either."

"You don't?" asked Candlelight. "I thought you lived here."

"No," answered Bramble, "we lived far from here in a place called the Bramble Thicket. Would you like to hear our story?"

Candlelight nodded. She was very curious about these rainbow-colored bears and where they came from.

"And we found you!" shouted Kandoo, as he danced a jig!

18

CHAPTER 4

Sassafras Cups

A warm, playful breeze rippled the leaves of the forest, then dashed down to swirl the dust of the path. The sunlight sparkled on the dancing water of the brook. Rough was very busy. He was doing something with the leaves of a scrubby little tree.

Candlelight was curious. "What are you doing Rough?" she asked.

"I am going to fetch you a drink of water," Rough said as he kept on working.

"We don't have any cups," Candlelight reminded him.

The bears laughed. "We have plenty of cups and plates too!" they said.

"Where?" she asked.

"On the Sassafras tree!" shouted the bears as they danced around her chair.

"On the what tree?" asked the amazed girl.

"Sassafras! It is a wonderful tree," explained Thicket, "from the root you can make lovely tea and from the leaves you can make beautiful green dishes!"

19

Whisper walked over to the scrubby little tree where Rough was still working. "This is a sassafras tree," she explained. "It has three kinds of leaves. One leaf looks like a plate." She picked a large oval leaf and laid it in Candlelight's lap. "The next leaf looks like a mitten." Whisper chose a large 'mitten leaf' and gave it to her friend.

"They are very soft," said Candlelight as she touched the leaves.

Rough picked one more leaf. "This one looks almost like a hand. It has three fingers," he said.

"But these are leaves, not cups," Candlelight reminded him.

"Not yet," said Thicket, "Look at Rough!" Rough folded the leaf that looked like a hand. One finger folded over the other, a small twig held the folded leaf together. Candlelight was amazed to see a tiny green cup! She watched as the 'mitten leaf' was made into a bowl. The large oval leaf made an excellent plate.

"It's like magic!" Her blue eyes could hardly believe what they saw.

Rough finished making a 'cup' for everyone. He was very proud of his work.

Ramble showed his friends a tiny spring of water gurgling from the rocks. Everyone enjoyed drinking the cool, delicious spring water from their little sassafras cups. But Candlelight enjoyed it the most!

"Bramble, you promised to tell me the story of how the Bramble Thicket Bears came to this forest," said Candlelight as she smiled at the little bear. "Please tell me now."

Bramble climbed up on an old stump. He sat down and began to swing his feet back and forth as he told the story.

"We are Bramble Thicket Bears. We lived far from here in a warm, sunny place."

"Where was this place?" asked Candlelight.

"It was at the end of the rainbow," answered Thicket, who sat near a bunch of daisies. The daisies seemed to nod in agreement as Thicket continued.

"Yes, the end of the rainbow, that's why we are the colors of the rainbow."

"Is there a pot of gold at the end of the rainbow?" asked the girl.

Bramble laughed. "The true gold is having a safe place to live, friends and contentment. Finding a pot of money at the end of a rainbow is impossible. Honesty, love, and work are the real riches! We know this is true, because one day we lost everything else."

Kandoo, who had been sitting quietly, spoke next. "The trouble started when a huge black thunder storm came to our valley. The lightning flashed and the thunder crashed and boomed . . ." Kandoo paused.

Bramble continued, "We had never seen a storm so terrible or strange! The rain never came, just the lightning and thunder." He hopped down off the stump and paced back and forth.

"What happened?" asked Candlelight.

"A fire started," answered Bramble, "a huge, raging forest fire. We hid in a cave, high up on the mountain. Many men and women came to put out the fire. They worked for many days. We were safe on the rocky mountain, but our home was destroyed. At last the fire was out, but everything was black. Even the little brook was black with soot and ashes." Bramble paused. Everyone was quiet for a moment as they thought of all that had happened.

"Tell her about Bucky," prompted Whisper.

"I can tell the story of Bucky," Ramble offered as he adjusted his yellow neckerchief.

"Indeed, that was a sad day, but we were safe and thankful to be together. We decided to travel across the mountain to find a new home. Early the next morning, while the fog lay in thick blankets over the valley, we began our journey. Many other forest animals were traveling across the mountain. We had gone only a short distance when we met a friend of mine. His name is Bucky."

Kandoo couldn't be still any longer. "Bucky is the biggest and best deer you ever saw! He gave me a ride!"

"Yes, Bucky is a fine friend," laughed Ramble. "He and his family carried us across the mountains to this valley."

"We were looking for a new place to live when we found you!" declared Rough.

24

Candlelight nodded at the bears as sunbeams brightened the sky, "I'm glad you found me. Something good can come out of every trouble."

Bramble agreed, "When trouble comes, we have to choose to do our best. Bramble Thicket Bears never give up! We always try to make things better and do our best to help others."

"How can we help Candlelight?" asked Whisper.

"She is too big for us to keep moving," observed Helper.

"And she eats a lot of berries!" said Tumble.

"We must find someone to help us," Bramble said as he sat down to think.

Bramble sat down to think.

CHAPTER 5

A Way To Help

Eight beautiful little bears gathered around Candlelight. They were all thinking of ways to help their new friend. Ramble paced back and forth. He walked to the rock, then back to the stump. Back and forth he went! Suddenly, he stopped.

"I know what to do! We will find Jonathan!" he exclaimed.

"Who is Jonathan?" asked Rough as he lay flopped in the grass. (Sometimes Rough did his best thinking while he rested!)

Ramble answered, "Jonathan is the best woodsman in the woods. He lives up on Heatherberry Mountain, and best of all, he is my friend!"

Thicket hugged Ramble. "You have solved our problem! When you need help friends are the place to find it," she said.

Candlelight was just as excited as the bears! She laughed as she watched Ramble unload the knapsack he always carried. Many wonderful things were in the knapsack. Three beautiful stones, a lovely leaf, one piece of bright blue ribbon, an old key, a whistle, a map and a sleeping bag were placed on the grass.

Rough and Tumble shook the sleeping bag, then rolled it up. Kandoo blew the whistle while Ramble looked at his map.

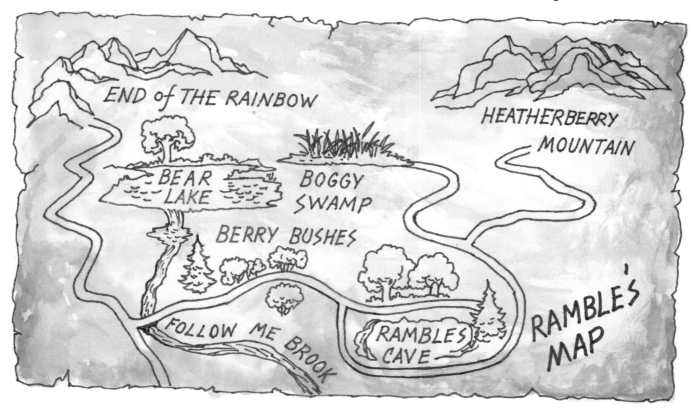

"Look," he said, "this is the valley where we are. (He pointed to a mark on the map.) And there is Heatherberry Mountain. I will go to Heatherberry Mountain and find Jonathan."

"May I go too?" asked Kandoo.

"No, not this time, you must stay here and help care for our new friend," he answered.

Ramble gathered up his treasures. He presented the blue ribbon to Candlelight. She thanked him and offered to keep his extra treasures in her pocket. The sleeping bag, map and old key were tucked back into the knapsack. Ramble hung the whistle around his neck, and Candlelight put his extra treasures into her pocket. Then the little orange woods-bear shouldered his knapsack, said goodbye to his friends and disappeared into the forest.

Bramble gave everyone a job to do. They were all very busy! Candlelight gave her handkerchief to Whisper and Thicket. They washed it carefully in the stream and laid it on the grass to dry. As soon as it was dry, Whisper and Thicket carried the handkerchief to the berry bushes and began to fill it with sweet, juicy blueberries. They filled it so full, they had to drag it back!

Rough and Tumble climbed a big hickory nut tree and jumped up and down on the branches. Nuts fell to the ground. Bramble and Kandoo hurried to gather the nuts. As Bramble bent over, one large hickory nut fell from the tree.

Kerplunk!

It hit smack on his seat!

"Ouch!" hollered Bramble, as he jumped. "Be careful where you drop those things!" Everyone laughed to see him hopping about.

Helper began to search for a large flat rock. He said to himself, "Just one flat rock . . . a large flat rock . . ." At last, he found just the right one! He cleared away the grass and brushed it off with his paw. Helper carried the rock to the hickory nut tree and dropped it on the grass. Then he went to find a small, heavy stone. When he found a stone he liked, he gathered a few hickory nuts.

"What are you doing?" asked Candlelight.

"I'm making a nut-cracker!" puffed Helper as he placed a nut on the large flat rock and smashed it with the heavy stone. **Smash! Crack! Smash! Crack!** Helper broke open all the nuts.

Nuts, berries, and spring water served on Sassafras dishes would make a delicious meal.

"I am hungry," said Rough as he rubbed his tummy.

"Me too!" agreed Thicket.

"Let's eat!" shouted Kandoo.

"I wonder where Ramble is?" said Whisper.

The others were wondering the same thing.

Nuts, berries and spring water served in sassafras dishes.

CHAPTER 6

Beartherapy

Candlelight's wheelchair fascinated Helper. He began to inspect it. (*Inspect* means to look at something very carefully.)

"Most amazing! **Most** amazing! A chair with two large wheels and two small wheels. Hmmm, it even has a floor for your feet. There are armrests for your arms. Yes, and this is the brake you showed us before."

"Candlelight, do you like the wheelchair?" asked Thicket.

"Well, yes and no," answered the young lady.

"Why 'yes?'" asked Bramble.

"It does help me move around, so I don't have to be in bed," answered Candlelight.

"Why 'no?'" asked Helper.

"Because I want to walk again. Sometimes it is very hard not being able to walk," replied the beautiful girl. "The doctor said I can learn to walk again, but I must have someone to help me."

"We can help!" the bears all spoke at one time.

"What should we do?" asked Thicket.

Candlelight was delighted. She laughed as she answered, "First, my feet and legs must be rubbed. Rubbing will keep them healthy. Second, my legs must be lifted up and down. Lifting will make my legs strong. After that, I must try to push my feet against the nurse's hands."

"Is this there-a-pea?" asked Kandoo.

"Yes," laughed the girl, "Exercise is therapy. Therapy is a strange word, but it is not too hard to say."

Kandoo tried several times to say the new word. He finally said therapy properly.

"We must take off your shoes and socks so we can rub your feet," said Thicket.

There was a lot of tugging, pushing, and giggling, but at last the little bears succeeded in removing Candelight's shoes and socks. Rough and Tumble each carried a shoe up to RAMBLE'S CAVE and set them side by side on the stone floor.

The bears began to rub her feet and legs. Thicket rubbed one foot while Whisper rubbed the other. Bramble began rubbing her right leg. It seemed like the colorful bears were everywhere! When it was Kandoo's turn, he climbed up on Candlelight's lap and slid down her left leg!

Swoosh!   Zoom!

He slid so fast he went flying right off her toes! Candlelight laughed until tears filled her bright blue eyes!

Bramble helped Kandoo up and brushed him off. "Come help us lift her legs up and down," he said. The seven little bears tried to lift Candlelight's legs, but they could not do that job!

"This is a hard problem," said Rough. "We need to think of another way to help her exercise."

Bramble agreed as he rested on the grass. Everyone was quiet. (They were thinking, and quiet helps thinking!) Suddenly, Bramble remembered the grape vines they had used.

"We can use the vines!" he exclaimed. "We will put the vine over a tree branch. Then we will tie one end to her leg, and we can pull on the other end. Her leg will go up and down!"

"**Hurrah! Whoopee!**" shouted the bears as Candlelight laughed and clapped her hands.

A large, low branch of the hickory nut tree was chosen. Rough tied one end of the vine around his tummy. He began to climb the tree. Up! Up! Up! he climbed. The others watched as he scooted out onto the big branch. Bramble held onto the other end of the vine. Carefully, Rough untied the vine from his tummy and let his end down to Helper. Lower, lower, lower went the vine. Bramble was standing on the tips of his toes, holding onto his end of the vine by his paw-tips.

"Rough, the vine is not long enough! I can not reach it! We have to tie another vine to this one," Helper called.

Kandoo brought another vine to Bramble, who was still holding on with his paw-tips. Tumble climbed up on Bramble's shoulders and tied the two vines together. They tested the knot. It was strong and tight. Once again, Rough lowered the vine. At last, it touched the ground.

"Now," said Bramble, "we must move Candlelight under the branch."

Rough climbed down from the tree, and together, the bears pushed the wheelchair over to the hickory nut tree. The soft vine was tied around Candlelight's right ankle. All the bears pulled on the other end of the vine. Up went the leg!

Helper held up his paw. "Stop!" he shouted. "Easy now . . . let it down easy."

"It works!" yelled Kandoo as he grabbed Thicket and danced a happy jig!

"Yes! Yes! It works!" declared Whisper, as she twirled in the bright sunshine.

Candlelight clapped her hands.

Up and Down! Up and Down! Twenty times the bears made the leg go up and down. Twenty more times for the other leg.

"This is work!" exclaimed Kandoo as he flopped in the grass.

Thicket agreed. "We should wait until tomorrow to do any more therapy!" she puffed.

"I hope Ramble comes back soon . . . he likes to work!" said Rough as he helped untie Candlelight's foot.

The group of friends settled down for the night. Stars twinkled overhead and the murmur of the pines sung them to sleep.

And so it was, when morning arrived, the bears were once again ready to work on "there-a-pea" (as Kandoo would say!). Bramble Thicket Bears may be small, but they are very inventive. They love to think of new ways to solve problems. This particular morning brought a new problem to solve.

Candlelight needed a way to push her feet against something. Helper soon had an idea! First, the bears put another vine across the low branch of the hickory nut tree. Candlelight was again moved beneath the branch.

"We will tie one vine to each leg!" said Helper.

When the vines were tied in place, Bramble and Tumble pulled on the first vine. Up went Candlelight's right leg! Rough, Kandoo, and Thicket pulled on the second vine. Up went Candlelight's left leg! Both her legs stuck out straight. Helper and Whisper hurried to tie the loose ends around a nearby log.

"What are you doing?" Candlelight asked, as Helper backed up to her feet.

"Push your feet against me," Helper said. "Come on, PUSH!" he encouraged, but . . . NOTHING . . . happened.

Candlelight gripped the arms of the wheelchair and concentrated (*concentrate* means to think very hard). "I'm trying," she said, but NOTHING happened.

Helper pushed his back against Candlelight's feet one more time. This time, he pushed too hard. Her knees bent . . .

KERPLOP!
Helper fell backward and landed on his seat.

"I'm sorry, I really was trying," giggled Candlelight.

Helper looked so funny she could not help laughing! The other bears laughed, and Helper grinned as he brushed himself off.

"I guess I tried too hard!" he said.

Thicket spoke between giggles, "Let me try this time." Carefully, she backed up to Candlelight's feet and snuggled her back against them. "Now, tickle me with your toes," she encouraged. "Just tickle me a little bit."

Candlelight tried very hard, but NOTHING happened. She was very disappointed.

"We'll do twenty leg-ups, then try again later," Bramble said as he reached up to pat Candlelight's hand. "Don't be discouraged! Success comes when you keep trying, even when you don't know why to try. Courage makes you keep trying."

The two vines were untied from the log, and once again the bears pulled on the vines. Up . . . Down, Up . . . Down, Up . . . Down. (It was a funny sight to see, sometimes one leg was up and sometimes both legs were up!) When the leg-ups were finished, the seven bears asked Candlelight to play a game. Their favorite game was hide-and-seek.

Candlelight closed her eyes and counted out loud while the bears hid, "One, Two, Three, Four, Five, Six, Seven, Eight, Nine, **TEN!**" She opened her blue eyes and tried to guess where each bear was hiding. Whisper and Kandoo were easy to discover because they giggled so much!

The bears asked Candlelight to play a game.

CHAPTER 7

Ramble's Journey

Ramble's journey took him over some very rugged country. Tall mountains rose on each side of the valley. Water rushed along in the mountain streams. Sometimes he had to climb up huge rocks! When the trail was clear, Ramble hurried as fast as his short little legs would carry him. When the path split, Ramble looked at his map. He figured out the best way to Heatherberry Mountain, then he built a pile of white stones beside the path. The white stones would help him find the way back.

On and on he hiked. A herd of deer passed by on the other side of the meadow. Later, he saw three baby raccoons playing in a hollow log while their mother napped nearby. A flock of geese flew high overhead, honking as though to say hello.

Ramble waved to them and kept on hiking. Up the hills and across the meadows he traveled, higher and higher up Heatherberry Mountain.

It was early evening when he saw Jonathan's cabin. What a welcome sight for a tired little bear!

The cabin sat in a lovely clearing, but between Ramble and the cabin was a D-e-e-p, **ragged** ravine. The ravine had steep rock walls on both sides. A mountain stream raced along the bottom. Ramble threw a rock into the ravine.

DOWN . . .

   DOWN . . .

      DOWN . . . it fell . . .

         SPLASH into the stream!

"How can I ever cross this ravine?" he wondered out loud.

When faced with a problem, the best thing to do is sit down and think about it, and that is exactly what Ramble did.

Should he try to swing across on a vine? "No, that is dangerous," he said to a curious squirrel. "Maybe I should cross on that old mossy log that is lying across the ravine."

Ramble went over to the moss covered log. He picked up a long stick and poked the log. He poked again and a **HUGE** chunk of bark fell off. It twirled as it fell . . .

DOWN . . .

DOWN . . .

DOWN . . . SPLAT!

The bark hit the water and began a swift journey down the stream. Ramble backed away from the old rotten log. He could not cross the ravine here! He sat down to think. There must be a safe way to cross over the deep ravine!

"Ramble, is that you?" someone called.

The little bear looked up. "Jonathan!" he shouted, "Is that You?"

"Yes, it's me," laughed the strong, slender young man. "What are you doing over there?"

"I am thinking of a way to cross this ravine," hollered Ramble over the sound of rushing water.

Jonathan pointed. "Go that way, there is a bridge!"

Ramble hurried along the edge of the ravine until he saw the bridge. What a strange bridge! It was made of rope and thick wooden boards.

"Is this bridge really safe?" asked Ramble in his very loudest voice.

"Oh yes indeed, I am sure the bridge is safe! I just crossed it this morning," answered Jonathan.

Ramble knew he could trust his friend, so he bravely stepped out on the bridge. It began to move, for it was a hanging bridge with a stout rope on each side. The little bear clutched the rope on the right side, but he could not reach the rope on the other side. One step . . . then another. Slowly he moved out onto the bridge. With every step the bridge moved up and down. Just as he was half way across, a playful breeze whisked its way up through the ravine. The bridge began to sway back and forth! Ramble stood very still and held on very tight.

"You are doing fine," encouraged Jonathan.

"I am frightened!" Rambled called in a nervous voice.

Jonathan knew Ramble could cross the bridge.

"You are stuck because you are not moving! Breathe deep and look at me. Take one step," he again encouraged his friend.

Ramble took a deep breath and bravely stuck his foot out and took one step . . . one more step . . . one step more, then he took three quick steps! He took deep breaths of air and gripped the guide rope with his paw as he continued across the swinging bridge. At last he was beside Jonathan!

"That is something I never did before!" Ramble said as he smiled proudly. "Thank you for helping me!"

"Crossing the bridge is fun once you learn how! You did very well!" his friend said. "Now what brings you here to see me?"

Ramble pointed to a nearby log as he took off his knapsack, "Let's sit down, and I will tell you."

Ramble told Jonathan about the forest fire and the journey over the mountain. He told how the Bramble Thicket Bears were searching for a new home. And how the bears had found Candlelight! He finished his story and said, "We need your help!"

The young man was amazed by all that had happened. "Wait here while I run to my cabin and get my knapsack. We will have to hurry back to your friends." Jonathan ran swiftly up the meadow path to his cabin.

Ramble, being a wise little bear, but a very tuckered-out one, (*tuckered-out* means very tired and sleepy) flopped down in the warm sunlight and dozed off for a short nap. (Naps are very helpful! They give you energy.)

A gentle hand shook the small orange bear. "Wake up little fellow, I'm back," said Jonathan softly.

Ramble mumbled something but did not wake up. Jonathan smiled, then reached down and picked up the tuckered-out messenger and Ramble's little knapsack. Ramble continued to sleep as Jonathan cradled him in his strong arm. Jonathan hurried across the bridge and began to jog down the mountain path.

It was dark when the young woodsman stopped to rest for the night. Ramble kept on sleeping as he was tucked under the corner of his friend's blanket.

The tuckered-out messenger continued to sleep.

CHAPTER 8

Tiggles

Birds sang a bright hello to the travelers as the morning sun chased away the shadows. Ramble crawled out from under the blanket and looked around.

"Jonathan look! There is a 'rainbow-catcher!'" he exclaimed.

His friend rubbed his eyes as he sat up. "A 'rainbow-catcher?' What is a 'rainbow-catcher?'" he asked.

Ramble pointed to a delicate spider web covered in dew. The sunlight danced across the dew drops painting them the colors of the rainbow.

"<u>That</u> is a 'rainbow-catcher!'" he stated. "See, it catches rainbows while you sleep!"

Jonathan laughed. "It looks like a spider web covered with dew to me," he said.

"Not if you are a Bramble Thicket Bear! If you are a Bramble Thicket Bear, <u>that</u> is a 'rainbow-catcher,'" Ramble declared with his paws on his hips.

Jonathan thought for a moment as he watched the glistening web, "I think you are right Ramble! It is indeed a 'rainbow-catcher.' I can see the rainbow colors sparkling in the dew!"

The two friends enjoyed blackberry jam sandwiches for breakfast. When they finished, Jonathan packed everything back into his big knapsack. Together they started down the mountain path. The long legs of one, made it hard for the short legs of the other to keep up. Jonathan stopped and waited for the little bear. Ramble was quite out of breath.

"Your legs are getting longer every minute," he gasped as he plunked down in the path.

Jonathan looked down at his legs. "I don't think they're getting longer," he said, "but they do seem to want to go pretty fast. Since you are the messenger and scout, how about climbing on top of my knapsack? Up there you can see much more."

Ramble thought for a minute. "I can walk all the way," he declared stoutly, "but if it would help you, I will sit up there."

Jonathan smiled as he knelt beside the brave little bear. "Climb aboard," he said cheerfully as he gave Ramble a boost.

Up to the top of the big knapsack Ramble climbed. He sat down between the shoulder straps and held on tight.

"Ready?" asked Jonathan.

"Ready," answered the scout.

The young woodsman began an easy run. Ramble held on tight! He had never gone so fast! Trees seemed to rush by. The ground was a long way down. His yellow neckerchief fluttered in the wind. At last, Jonathan stopped.

"Which way should we go?" he asked.

The path they had followed split; one led over a brook, the other across a meadow. Ramble looked at both paths, then he saw the pile of white stones he had placed as a marker.

"There's my marker!" he said, pointing to the stones. "We take this path and go over the brook."

Across the brook . . . on and on they quickly traveled until they reached a small hill covered with fir trees. Jonathan stopped.

"Listen," whispered Jonathan, "do you hear something?"

Ramble listened. "Yes, I hear a lot of hollering. I wonder what all the noise is about?"

Jonathan knelt down and leaned back until his big knapsack almost touched the ground. Quickly the Scout climbed down. "Let's move forward very quietly and see," Jonathan said.

Together they crawled forward and peered through some bushes. They saw a most amazing sight!

Vines hung from a large hickory nut tree. A young lady in a wheelchair was beneath the tree, and the vines were tied to her legs. Little bears of all colors tumbled and shouted around her!

In a flash, Jonathan was kneeling beside the girl and looking into her blue eyes. "Are you all right?" he asked.

She nodded yes, but tears flowed down her cheeks.

"Quiet!" commanded the young man.

Instantly, there was silence. The bears froze in place, wondering what was happening.

Jonathan took the young lady's hand and asked again, "Are you all right? Why are you crying?"

"Oh yes, I'm fine! We are just so happy!" she said, smiling through her tears.

All the excited little bears started to explain at one time!

"She moved her toes!"

"This is 'there-a-pea!'"

"We are helping!"

They were all talking at once as they danced around.

"Be still!" Ramble said quite loudly.

Everyone became quiet once more.

"Thank you, Ramble," said Jonathan. "Now, one at a time. What is going on here? Why is this young lady tied to that branch?"

Candlelight explained, "The bears are helping me exercise!"

Kandoo said, "Let's show them."

And that is exactly what they did.

Bramble, Helper and Rough pulled on one of the vines. Up went Candlelight's leg! They lowered her leg carefully.

"These are called 'leg-ups,'" said Thicket.

"We helped her try 'push-mes' too," added Whisper.

"But I cannot do the 'push-mes' yet," said Candlelight.

Whisper climbed up on Candlelight's lap. "But she moved her toes! That's why we were so excited! We will show you!"

The bears took their places. Up went Candlelight's legs. Thicket backed up to her feet.

"Think hard!" encouraged Rough.

"You can do it!" said Kandoo.

Candlelight concentrated (thought very hard!). She wanted to wiggle her toes. She tried . . . one toe wiggled! She tried again . . . the big toe on her other foot moved!

"See, she did it!" shouted the bears.

Tears gleamed in the girl's eyes. "They wiggled! Look! My toes wiggled!" she said.

"That is wonderful!" said Jonathan with a big smile, as he gently untied her legs. "Now we need to make plans to get you home."

"Are you Jonathan?" asked Rough.

"Yes, I am Jonathan," he answered. "Ramble told me you needed help. We will send a message to my father at his farm."

"How?" asked Kandoo. "Can your father hear us if we shout?"

"No," laughed Jonathan, "Tiggles will carry the message."

He opened his knapsack and took out a small wooden box with holes in the sides and a little wire door on the front. Inside the box was a beautiful, soft pigeon. The bears were very curious. They had never seen a carrier pigeon before. The little rainbow-colored bears sat quietly as Jonathan wrote a message on a tiny piece of paper and rolled it up. He took Tiggles from the box and carefully tied the message to her right leg.

"Fly Tiggles!" Jonathan said as he gently tossed the little pigeon into the air.

Tiggles flapped her wings and circled high above their heads, and then headed straight for the farm. The Bramble Thicket Bears became sad, knowing Candlelight would soon go home.

Inside the box was a beautiful, soft pigeon.

CHAPTER 9

The Invitation

Candlelight called Thicket, and spoke very quietly to her little friend. Thicket listened and began to laugh!

"What is so wonderful?" asked Rough in a low growly voice.

The others wondered the same thing.

"Tell them, Thicket," the girl prompted.

The little burgundy bear looked at all the other bears as they gathered around the wheelchair. "Candlelight wants us to go home with her!" she announced.

"Home with Candlelight? Where would we live?" asked Bramble.

"My grandfather has gardens, meadows and a forest. I'm sure you could find a home," Candlelight answered.

The bears huddled together. They talked about this new idea. Bramble stepped forward and spoke, "We would like to go home with you. It would be fun to live close by!"

Jonathan had just finished putting Tiggles' box back into his knapsack, when Kandoo ran to tell him the news.

"We are going home with Candlelight!" Kandoo shouted.

Jonathan smiled as he looked down at the excited little bear. "Who is 'We?'" he asked.

Candlelight laughed, "We forgot to introduce everyone!"

The bears scurried back and forth until at last they stood in a straight line. One bright blue bear, one light blue bear, two green bears, one yellow bear, one orange bear, one pink bear, and one burgundy bear all stood in front of Jonathan.

Candlelight pointed to the bright blue bear, "This is Bramble. Next is Helper. The twins are Rough and Tumble. Kandoo is the yellow fellow, and you know Ramble. Whisper is the pretty pink bear, and Thicket is the burgundy bear."

The bears were very polite. They each said hello. Jonathan was pleased to meet Ramble's friends.

Kandoo pointed to Candlelight. "She is Candlelight," he said.

"I know," smiled Jonathan. "Now we must clean up the camp. Good campers always leave the campsite neat! Dad will be here soon with his horses and wagon. We must be ready to meet him."

The vine ropes were taken down and the extra nuts set beneath the tree for some lucky squirrel. The bears brushed the nut shells off the flat rock and brought Candlelight's shoes to her. Jonathan began to put her shoes on her feet.

"I feel like a real princess," Candlelight said as Jonathan knelt before her.

"And you're just as beautiful," said Jonathan. "We shall use your chair as a coach! Some of the bears will be the horses. Ramble, you shall be the coachman!" He lifted him up onto Candlelight's lap.

The other bears lined up in front of the wheelchair. There were no vines to pull this time! The bears pretended to be horses.

"Tally ho!" shouted Ramble, and off they went.

Helper looked at Bramble as they trotted in front of the wheelchair. "Pretending is a lot easier than pulling!" he laughed.

Jonathan made the wheelchair roll easily over the ground until they came to the old dirt road. Soon Jonathan's dad arrived driving a large red wagon pulled by two Clydesdale horses.

Jonathan's father was very kind and made sure everyone was comfortable for the long ride.

It was twilight when the wagon arrived with its passengers at Candlelight's home. Grandfather's search for Candlelight was over. He was very pleased to see his lovely granddaughter and to meet her new friends. He invited Jonathan and the bears to stay the night in the beautiful stone mansion. As Grandfather led the way inside, he smiled at the excited bears.

"I never slept in a house before," whispered Thicket.

Grandfather heard her. "We have a special place I think you will like," he said. "But first, Candlelight will show you our home." Then with wink and nod he added, "I will meet you in the conservatory."

The bears were thrilled to see the beautiful rooms and the sparkling chandeliers. Whisper and Thicket loved the pictures of flowers woven in the carpets. Bramble was quite fascinated by the old pictures, especially one of a little boy holding a bear.

On down the Great Hall went the adventurers, peeking into the rooms on each side. At last, they came to a large door made of oak. Candlelight pressed a button and the door swung silently open.

"IT'S A FOREST!" shouted Tumble.

"It's lovely!" sighed Whisper.

"It's your room for tonight!" said Candlelight. "We call it the conservatory. It's a place where we grow special plants."

Grandfather entered the room. "Do you like it?" he asked.

"Oh yes!" chorused the bears.

"Tumble and I think it is a Forest!" explained Kandoo.

"I thought you might like your very own forest," said Grandfather with a chuckle. "I have asked the servants to set out a few things for you . . . here they are!"

He pulled back a large fern and showed the bears a small, round table with chairs just the right size! On the table were bowls of berries, nuts and a plate of warm nutmeg cookies. Mugs of milk were at each place.

Grandfather set a large package on the floor, and whispered something to Candlelight. She called the bears.

"This is for you," she said, pointing to the package.

The bears hurried over to her wheelchair. Each bear said, "Thank you." Then Jonathan turned Candlelight's wheelchair around and followed Grandfather into the hall. The bears called, "Good Night," as he shut the door.

The Bramble Thicket Bears were very curious about the big package that lay in front of Bramble. It was wrapped in brown paper and tied with a string!

"What is it?" asked Thicket.

"It's a big package," said Ramble.

"OPEN IT!" shouted Kandoo.

Bramble untied the string which was around the brown paper bundle. Then he opened the paper. Inside were eight soft blankets, each one was a different color.

"There's one for each of us!" said Helper.

"We will take turns choosing," said Bramble. "Kandoo will start."

Kandoo chose the purple and white striped blanket. Whisper liked the lavender one. Thicket had a hard time making up her mind, but finally picked the light green blanket. Rough and Tumble were next. Rough chose the blanket with orange and blue stripes, and Tumble picked the blue blanket. Helper liked the gray one and Ramble took the brown. Bramble looked at the last blanket. It was yellow, and it was the one he liked best!

Of course, all the bears had to wrap up in their blankets and dance around. They looked very silly!

Thicket reminded them of the good things to eat. Eight hungry bears soon finished their supper!

High above the glass roof, the stars twinkled in the night sky as the bears, wrapped in their blankets, settled down for the night.

Whisper spoke softly, "I'm glad we did our best to help Candlelight."

"God bless our new friends!" said Helper as he snuggled under his new blanket.

The bears had lots of fun whirling and dancing.

"I have a mystery for you and your friends to solve."

CHAPTER 10
Grandfather's Mystery

The warm spring morning brought the fragrance of lilacs to Candlelight as she wheeled herself out onto the stone terrace. Dew sparkled on the grass as the birds began to sing. The door behind her opened, and Grandfather came quietly forward.

"Morning is special isn't it?" he asked.

The girl nodded, not wanting to break the spell of the world awakening. Grayish-green shapes began to show their true colors as the sun peeked over the hills. Mist rose from the dewy grass. In a nearby urn, yellow roses glistened in the rising sun.

Several moments passed in silence before Grandfather spoke again, "I must leave on business. Mrs. Pennywhistle is here and has promised to stay with you. Jonathan has also agreed to stay until I return tomorrow."

Mrs. Pennywhistle was the housekeeper for Grandfather. Candlelight loved her, and Mrs. Pennywhistle loved Candlelight just as if she was her own daughter.

Grandfather continued, "The police have caught those that took you away. They will not bother you again. I'll be back soon."

"I will miss you," said Candlelight.

"You will be too busy to miss me. I have a mystery for you and your friends to solve!" Grandfather placed a large, beautiful old key in her hands as he spoke.

"What is this for?" asked the girl as she examined the key. (*Examine* means to look very closely at something.)

"That is the mystery! If you have a key maybe you should look for a lock," Grandfather's eyes twinkled as he spoke.

Candlelight thought for a minute, "If I want to find a lock maybe I should find a door," she said, "or a box."

"Or perhaps a hidden garden," Grandfather suggested as he kissed Candlelight good-bye. "Jonathan and the bears will be glad to help you look."

While Candlelight and Grandfather were talking, the sun painted rainbows through the glass roof and walls of the conservatory. The dancing light awakened Kandoo, who poked Bramble and asked, "Are we in heaven?"

Bramble peeked out from underneath his yellow blanket, "No, it's not heaven! And I'm hungry!" he said.

As Whisper straightened her ribbons, she noticed the little table beside the fern. It was set for breakfast. Little bowls of hot oatmeal were at each place. A large bowl of berries was in the center of the table. Mugs of cold fresh milk sat waiting for the bears.

"This is magic," said Thicket. "We didn't have to find our own breakfast!"

"This is wonderful," agreed Helper as he stretched, "but we must find our own place to live. Life is more than beauty and magic. One must work and be useful."

"I believe that if you never had to think and work, you would forget how!" Bramble declared.

"We need our own place so we will be responsible," said Ramble.

"Let's hurry and go outside so we can find Candlelight and Jonathan," Kandoo urged the others as he tasted his oatmeal.

As soon as they finished breakfast, the bears pushed the chairs up close to the table. They folded their blankets and stacked them under a fern. When everything was neat, they scampered outside to find their friends.

"Hello!" called Candlelight as the bears raced toward her.

"We were coming to wake you up!" laughed Jonathan. "We have a mystery to solve."

"A mystery?" asked the bears as they tumbled over each other in their hurry to reach Candlelight and Jonathan.

Candlelight showed the Bramble Thicket Bears the old key.

"Grandfather gave me this key which opens the door of a hidden garden," she explained. "Will you help us find the garden?"

"Yes! Yes!" chorused the happy bears. "We will help you!"

All morning long the rainbow colored bears dashed here and there among the shrubs and flowers and paths of Grandfather's estate. They were looking for a door which had a lock for the old key.

It was almost noon when Bramble and Thicket ambled down an old stone path. Bramble was tossing a pebble into the air and trying to catch it.

"Oops! I missed!" he said, and as he bent over to pick up his pebble, he saw something strange. "Look!" he said to Thicket, "there's a red marble in this stone!"

"Let's see if there are any more," said Thicket.

"Here's another one, it's yellow!" shouted Bramble

"I see a blue marble in that stone!" Thicket exclaimed.

The two bears followed the path of marbles and stones. At last they came to a high stone wall, and in the wall were two big, heavy, old wooden doors.

"Where is the lock?" asked Thicket.

"There is no lock," said Bramble. "Come on, we should find the others."

"We need Jonathan to open those doors," Thicket said, as she followed Bramble down the curving pathway.

They hurried across the rose garden and around the gold fish pond to the big oak tree. The other explorers were waiting in the shade of the big oak. They had found several beautiful gardens but no one had found a hidden garden.

"We found two old doors, without a lock," Bramble shouted.

Candlelight leaned forward. "Two doors?" she asked, "how did you find them? Where are these doors?"

"We followed the marbles in the old path," explained Thicket.

"Show us," the other bears said.

"Is it far?" asked Rough. "My feet are tired!"

Jonathan laughed. "Okay, who wants a ride?" he asked.

All the bears wanted to ride! They began to climb up on Candlelight's lap. "I can't hold all of you," she giggled.

Jonathan put Rough on one shoulder and Tumble on the other. Then he buttoned Kandoo inside his shirt so only Kandoo's head stuck out! Whisper and Thicket snuggled on Candlelight's lap. Helper sat on her feet, while Bramble straddled one arm of the chair and Ramble straddled the other. Bramble gave directions and off they went!

There was a lot of giggling and shouts of delight as the strange procession went around the gold fish pond and across the rose garden. Helper bounced off once, but he wasn't hurt. He climbed back on and held tight to Candlelight's shoe laces.

They arrived at the far side of the rose garden and followed the marbles to the old wooden doors. Jonathan and the bears all tugged on the doors. The doors sagged on the hinges and creaked as they swung open. Inside was a stone tunnel. The floor was paved with bricks, and at the other end of the tunnel stood a door with a small window in it. This door was **locked**!

The bears were silent as Candlelight gave Jonathan the key. The young man put the key in the lock. Click . . . Click . . . Click . . . the key worked! Slowly, he pushed open the door . . .

The young man put the key in the lock.

CHAPTER 11

A New Home

Beautiful flowers grew everywhere! There were flowers of all different colors; blue, pink, yellow, red, purple and peach!

"It looks like a rainbow!" said Whisper.

A path winding through the garden seemed to beckon the bears, calling them to come explore the newfound garden.

Helper hopped down from the wheelchair and pulled a branch out of the way. Jonathan pushed the wheelchair forward. They were surrounded by many beautiful flowers. Sun beams glistened on a waterfall which splashed its way down to a cool clear pond.

The eight little bears began to explore. They scrambled here and there peering into everything. Rough and Tumble found a spot of green grass. It was soft and thick and perfect for running, tumbling and jumping!

Tumble did a headstand. (He liked to stand on his head and see everything upside down!) Then he noticed something under a bushy shrub. Tumble flipped down on his tummy. Sure enough, there was something under that bush. It looked like a large square stone. He wiggled under the bush, then he wiggled some more. At last he bumped his head on a cold stone wall. Tumble was puzzled. What was a stone wall doing here in the middle of the garden? Carefully he pushed the branches aside and stood up. There looking right back at him was a dirty little bear!

Tumble plopped down quickly! Then he cautiously peeked over the stones again. There was that little dirty bear peeking back at him again! This was more than the little green bear could understand! He scrambled away and wiggled out from under the bush.

"Jonathan, Bramble, Helper . . . HELP!" shouted Tumble.

"What's the matter?" asked Jonathan as he came running.

"There's a strange little dirty bear behind that bush!" gasped Tumble, pointing his paw to the place he had been in the bushes.

Jonathan cautioned the bears to stay back as he investigated the bushes and whatever was behind them. It only took a moment for the young man to see what the problem was. He began to laugh so hard that he could not speak. The bears were very puzzled.

"What's so funny?" called Candlelight from where she sat.

Jonathan finally spoke through his laughter, "Tumble found a dirty little bear back here, and I know who that dirty little bear is!"

"Who?" asked Tumble, as he helped Jonathan come out from the bushes, "who is the dirty little bear?"

"YOU!" laughed Jonathan as he rolled in the grass holding his sides. "You are that dirty little bear!"

Tumble was puzzled. He really did not see what was so funny. Candlelight was puzzled too, "What do you mean?" she asked.

Jonathan sat up and tried not to laugh. "Tumble has found a little glass house. He saw his own reflection in the glass and thought it was a strange little bear. Actually, he was just looking at himself in a big mirror!"

Now all the bears began to laugh, even Tumble, who now felt better since he had discovered something new.

The afternoon sped by as the bears and their friends cleared the grass and bushes away from the glass house. At last, there before them stood a little octagon-shaped glass house. The little house was just the right size for the Bramble Thicket Bears!

Candlelight asked, "Could this be your new home?"

"Well," said Bramble as he looked around, "It is just the right size and we all love the garden. It would be wonderfully delightful to live here." The other bears agreed, although they did not use such big words! There was a lot of paw patting and hugging and good old celebrating that day.

Early the next morning the bears were hard at work when Jonathan pushed Candlelight's wheelchair into the garden. Candlelight was delighted to see stones placed neatly in rows along the path. Old leaves and twigs had been cleared from under the shrubs. The garden sparkled with new life and beauty.

"We brought you something," announced Candlelight as she held a large box on her lap.

Inside the box were all sorts of little buckets, sponges, cloths and brushes. There were two bars of soap too.

Jonathan filled the buckets with soapy water and set them on the stone walk. Soon he was busy lifting first one bear and then another high into the air as they scrubbed the little glass house. Candlelight was kept busy rinsing dirty sponges and placing them in the warm sunlight to dry. Helper and Tumble used the brushes to scrub the stone steps.

Mrs. Pennywhistle brought a good lunch of applemint tea, peanut butter sandwiches, and lemon drops. She was very proud of all the hard work that everyone was doing.

As the evening sun set in the west, it sent its last glimmering rays to shine on the little glass house.

"Oh," exclaimed Whisper.

"Look!" shouted Tumble as he turned a somersault.

"Can you believe it?" whispered Candlelight to her friends as they stood beside her in the glass house.

"It's a dancing Rainbow!" exclaimed Bramble as the rays of light cast hundreds of rainbows all around the room.

"It's magic!" said Kandoo excitedly.

"It's the Bramble Thicket," laughed Candlelight.

"Yes," the bears agreed happily. "It's the Bramble Thicket now and forever."

"It's the Bramble Thicket now and forever!"

CHAPTER 12

The Secret

Sparkles of color chased each other across the clean glass walls of the octagon-shaped house. Helper and Jonathan were very busy working. They were fastening a brass bell above the door of the glass house. They had just finished their work and had gone inside when the door opened and the bell tinkled merrily.

"Hello!" someone called.

"Hello, sir!" said Jonathan, as he shook hands with Grandfather.

Helper offered his paw. "We found your hidden garden, and Candlelight said we can live here forever! It's our new Bramble Thicket!" he explained as Grandfather shook his paw.

Grandfather laughed. "I am glad to hear that!" he said. "Please find Candlelight and have everyone come up to the terrace. I have some important news for you." Grandfather waved to them as he left the garden.

Helper and Jonathan hurried to find Candlelight and the bears. Soon they were all gathered at the stone terrace. Rough and Tumble played nearby on the swing which hung from a large beech tree. The other bears played leap-frog while they waited.

Soon Grandfather appeared carrying three packages. They were wrapped in pretty paper and tied with ribbon.

"I have some thank-you gifts for you," he said with a twinkle in his eye.

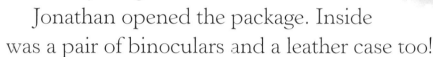

"Jonathan, this is for you. Thank you for bringing Candlelight back home." Grandfather shook hands with the young man.

Jonathan opened the package. Inside was a pair of binoculars and a leather case too!

"Thank you sir, they are wonderful!" he exclaimed.

Grandfather set the next package on the stone floor and said, "This is for all the Bramble Thicket Bears. Thank you for finding and taking care of Candlelight."

The bears began to unwrap the package. It was a bit difficult with paws, but they soon had the box open. A set of small china dishes lay in the open box.

Whisper counted the plates, "One, Two, Three, Four, Five, Six, Seven, Eight," she said as she touched each plate.

Rough counted the cups carefully, "One, Two, Three, Four, Five, Six, Seven, Eight," he said. "That's one for each of us!"

Eight bears remembered their manners and said "THANK YOU!"

Grandfather gave the last package to Candlelight.

"Thank you, Granddaughter, for being cheerful and never giving up hope," he spoke as he kissed the top of her head.

Candlelight opened her box as the bears peered curiously over the sides of the wheelchair. She took out the most beautiful blue gown she had ever seen. The soft fabric shimmered as she touched it.

Candlelight was puzzled, "Where will I ever wear this beautiful gown?" she asked.

"You will wear it to the Christmas Ball," said Grandfather.

"And I would like the honor of taking you to the ball," Jonathan said with a deep bow and a big smile. "Will you go with me?"

"Yes! Thank you! It will be wonderful to go to the Christmas Ball!" Candlelight laughed as she spoke.

The days flew past, and the months seemed to fly too! True to their promise, the bears helped Candlelight exercise every day. They made therapy fun! Her legs grew stronger and stronger. The bears and Candlelight kept this a secret!

At last, the night of the Christmas Ball arrived. Snow covered the gardens and reflected the lights which glowed from each window. Grandfather's guests soon arrived. The ladies wore lovely gowns and the men wore black tuxedos.

Laughter filled the crisp air and music filled the house. Candles and chandeliers spread their glistening light. Tables held delicious food and beautiful flowers.

The Bramble Thicket bears were everywhere, investigating everything until the guests began to arrive. Candlelight knew the bears would be shy among so many strangers.

"Come with me," she said, as she led them to the ballroom balcony. "See the big pots of ferns? Grandfather and I thought that you would like to hide among them and watch the Christmas Ball."

Just at that moment, they heard the front door-bell chime. Dong! dong! dong! The bears dashed into the hall, and arrived at the top of the stairs as the big front door opened. Jonathan entered. He removed his cape with a grand flourish that left the bears wide-eyed and dazzled. Then he shook hands with Grandfather. The bears stared. They could not believe their eyes! The young woodsman looked very handsome in his dark tuxedo.

"We have a secret!" Kandoo shouted down to his friend. In his excitement to reach Jonathan, he tipped over a large fern. The rest of his sentence was lost as the fern swayed and fell with a loud crash! Kandoo disappeared in a wild flurry of paws and leaves, but soon came up sputtering:

"L-L-L-Lo-o-o-k-k-k," he gasped, brushing a fern leaf from atop his nose. "Look at Candlelight! AAAAchoooo!" The little bear sneezed before dropping out of sight behind the fern.

Jonathan looked up at Candlelight as she stood at the top of the stairs, laughing at Kandoo.

The bears saw the look of amazement on Jonathan's face as Candlelight walked down the stairs. Yes, walked! All the hours of Beartherapy had brought success!

81

The bears watched as the beautiful girl and tall young man walked to the ballroom. The young couple paused and turned. Candlelight blew a kiss to the Bramble Thicket Bears. The music began and the couple glided away.

The rainbow colored bears watched from the balcony, until they fell asleep to dream of adventures to come.

The End